Mathematics
The Search for Law, Order, and Patterns in the Universe

By C.R. Lind

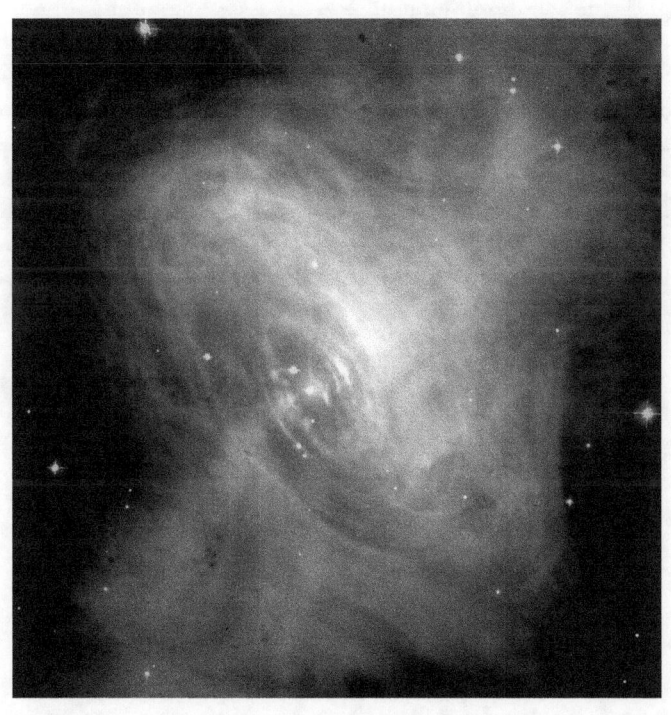

Table of Contents

Chapter 1

Law and Order, The foundation of Mathematics

The origin of consciousness...
The nature of designed objects...
The foundation of law and ethics...
The behavior of markets and the economy...
The source of order in our lives and in the universe...

All begin with an understanding of Law and Order. What are laws? Laws are rules of behavior. The single greatest achievement of mankind is the discovery that the entire universe, every speck of matter in it, is governed by laws... by rules of behavior. These laws create the order we see in the world around us.

What is order? Order is a predictable pattern of behavior.[1] This is where mathematics comes in. Mathematics has been called the science of patterns. It is the search for the underlying cause of the order...the patterns...we see in the world around us.

Reality has an underlying order to it. This order is created by natural laws. These laws are called natural because human beings did not create them. They are intrinsically

---

[1] The opposite of order is randomness. Randomness is essentially defined as a pattern or behavior which cannot be created by or described in terms of simpler underlying laws. This definition of randomness was discovered by IBM researcher Gregory Chaitin. This definition is part of his theory of information called algorithmic information theory. Randomness is not chaos, which is now a branch of mathematics which deals with laws that produce behavior that only LOOKS random. See chapter 5 for more information.

part of nature. These natural laws apply to everything and everyone in the universe, whether they are aware of them or not. Natural laws are discovered through the use of <u>REASON</u>. Reasoning can be used to discover natural laws precisely because reality has an underlying order to it.

Natural laws are discovered by using your intelligence (ability to reason) to perceive and recognize patterns in nature. New mathematics is discovered by using your ability to reason to perceive and recognize patterns in numbers.

There seem to be two types of laws. There are the laws that people discover (the natural laws) and laws that people create, called positive laws. Textbooks on politics, government, and legal systems contain the positive laws that people in a city or nation are abide by. Physics and chemistry textbooks contain the known natural laws that people have discovered to date. The mathematical formulas and symbols in these text books are but symbolic representations of the actual laws found in nature. They are the map of the territory.

Another name for a law or rule of behavior is an algorithm. An algorithm is essentially a step by step list of instructions. In mathematics textbooks, algorithms are called functions and in physics, algorithms are called natural laws.

Laws and rules produce order (predictable patterns of behavior) and not the other way around.

Both natural and positive laws create order and both types can be discovered by looking for patterns in the behavior of the world around you.

For example, if you noticed that everyday at 8:00 am, the employees of a restaurant arrived for work, you would guess that a rule or law of that restaurant is that employees arrive by 8:00 AM.

You discovered this law by noticing a pattern of behavior in nature or the world around you (the behavior of the restaurant employees in this example), and then taking a guess as to what the rule or law that produces this pattern might be.

Similarly, by noticing the pattern of the elliptical orbits of the planets around the sun, astronomers were able to discover the mathematical inverse square law of gravity.

When you pull the plug in a bathtub full of water, a small whirl pool forms around the drain. This behavior is produced by the laws of gravity and inertia acting on the water.

Order allows us to structure and plan our lives around future events.

If the owners of the restaurant decided to no longer have a rule governing when employees show up for work, chaos would quickly follow. Employees would simply show up whenever they felt like it. If the chefs showed up at 2:00 pm some days, and 10:00 am other days, and some days not at all, customers would not know what times or days they could expect service and the restaurant would quickly go out of business.

Order is what makes possible engineering, or designed objects.

If the laws of gravity suddenly and randomly changed from

moment to moment, virtually all of our technology (and probably the entire universe), would break down. It would not be possible to design an airplane, a building, or a bridge, if the weight of these objects changed randomly from moment to moment.

In summary, "laws CREATE ORDER or PATTERNS". Whenever you see a pattern, there is an underlying law (algorithm, function, or natural law) that has caused it.

The single greatest achievement of mankind is the discovery that the entire universe, every speck of matter in it, is governed by laws and rules of behavior.

Surprisingly, just because all of the atoms and subatomic particles in the universe are governed by natural laws, this does not necessarily mean that all phenomena in the universe are created by natural laws.

There are some phenomenon in the universe that, try as we might, quite simply have not yet been explained entirely in terms of matter and natural laws.

Coming in at number one is the phenomenon of mind, or consciousness.

How is it possible for consciousness, or a mind, to be created from unconscious, non-self aware matter and natural laws? Is it possible? A lot of AI researchers would say yes. But surprisingly, some very bright thinkers in the fields of AI and philosophy of mind are now saying no.

See the website of David Chalmers for a look at a leading philosopher of mind who is arguing that a form of dualism is in fact the only way to explain mind or consciousness. See my review of David Chalmer's book, "The conscious mind".

For a purely materialistic view of philosophy of mind, see Daniel Dennett (Consciousness Explained) and Douglas Hofstadter (Gödel, Escher, Bach, an eternal golden braid).

Coming in at number two is the nature of a designed object.

I spoke about engineering, or the art of creating designed objects, at the top of this paper. But what IS a designed object? Presently, if we want a designed object, we have to hire an intelligent designer...an engineer, to design and construct the object. Is this a pattern of design in general...that to get a designed object you need an intelligent designer? Or are matter and natural laws sufficient to explain all design (or "apparent design" if materialism is correct) in the universe?

In a sense, this question boils down to the question of the origin of mind. If mind turns out to be nothing more than the product of matter and natural laws, then all designed objects produced by "intelligent designers" must be the product of matter and natural laws as well.

The evolution debate is where you can find a lot of information on this topic. The leading proponent of design being purely a product of matter and natural laws is Richard Dawkins (See his books, the selfish gene and Climbing Mount Improbable).

The leading proponent of the necessity of an intelligent designer is William Dembski (see his books, No Free Lunch and Intelligent Design).

A lot of mathematicians these days are focused on the origin of order in the universe. Where does order come from? Not all phenomena exhibiting order in the universe are designed. The wave patterns formed in the sand dunes of a desert can be explained without recourse to an intelligent designer, but that might not be the case for all natural phenomena in the universe.

Presently, the only objects we can say with certainty are designed are the ones we ourselves have created. Everything else is still subject to debate.

Coming in at number three is the nature of ethics. Is human behavior the product of culture, or is there a natural moral law which guides our behavior and actions towards one another?

If tomorrow, all of the governments and laws in the world, all reference to law and government in the world contained in books and other media, and all memory of law and government were suddenly erased from existence, what would be the result on human behavior toward one another? Would moral and ethical behavior suddenly stop at once? I believe that the answer to this question is that no, it wouldn't. I do not believe that I personally would suddenly behave like an irrational, amoral animal or immoral person under such circumstances. If people behaved morally under such conditions, then could you say that their behavior was caused by an underlying moral law?

Again, in my opinion, this question boils down to philosophy of mind. If our minds are truly produced by matter and natural laws, then how is it even possible for an ethics to actually exist? To me ethics implies free will, or the ability to choose between different courses of action, some moral and some immoral. Free will implies a mind that is independent of the deterministic natural laws producing the behavior of a brain. (Interestingly enough, philosophers such as Daniel Dennett argue that even if mind is created and governed by matter and natural laws, free will is still possible.)

J. Budziszewski, author of Written on the Heart, the case for natural law, is a popular philosopher of the natural moral law.

See Alan M Dershowitz, author of Rights from Wrongs, for a look at a law professor who is trying to ground a theory of ethics while simultaneously denying the existence of a natural moral law.

Chapter 2

Randomness, Order, and Information

What is randomness?
What does it mean when someone says that a number or
event is random? Intuitively, a non random event is one that
follows a pattern.

In theory, tossing a coin into the air is supposed to be a
random event.

For example, which of the following is the result of tossing
a coin into the air 20 times and writing a 1 when the coin
lands heads and a 0 when the coin lands tails?

01110100001010001101

or

10101010101010101010

It is intuitively obvious that the first series of 1's and 0's
was a result of tossing a coin into the air 20 times, but not
the second. Why? Because the first result DOES NOT
follow a definite pattern, whereas the second one does. In
other words, we perceive the first string of 1's and 0's to be
random BECAUSE it does not follow a specific pattern.

As another example, imagine you have a friend go into
another room and toss a coin 20 times in the air. When your
friend returns from the room, they tell you that the result of
their tossing a coin 20 times into the air was
10101010101010101010. Would you believe them?
Probably not. Why, because this series of 1's and 0's too

closely matches an obvious pattern.

The amazing thing is, even though it is intuitively obvious that the first series of 1's and 0's is random, according to classical probability, BOTH of these results have EQUAL likelihood of occurring. According to probability theory, there are 2^{20} different permutations of a coin tossed 20 times into the air. Each of the above sequences of 1's and 0's represent just one of these 2^{20} permutations. So the likelihood of the first result occurring is $1/2^{20}$ and the likelihood of the second event occurring is also $1/2^{20}$.

So probability alone (or the unlikelihood of an event occurring) CANNOT be a measure of randomness. If this were not the case then you would have NO reason to believe that your friend had lied when they told you that the result of there 20 coin tosses was 10101010101010101010.

Where these 2 sequences of 1's and 0's DO DIFFER is that the **second one can be produced / generated by an algorithm**, whereas the first one cannot.

An algorithm is a step by step list of instructions. (For more information on algorithms and how they relate to computers, see chapter 7.)

An algorithm to print the sequence 10101010101010101010 is:

Write "10" ten times.

An algorithm written in Visual Basic to produce the sequence 10101010101010101010 is:

```
dim i as integer
for i = 1 to 10
```

```
print "10"
next i
```

On the other hand, an algorithm to produce the sequence
01110100001010001101 is Write
"01110100001010001101".

An algorithm written in Visual Basic to produce the
sequence 10101010101010101010 is:

```
print "01110100001010001101"
```

**Notice what happens when we extend the two sequences
of numbers from 20 to 40 digits:**

The algorithm to produce
1010101010101010101010101010101010101010

is: print "10" 20 times,

whereas the algorithm to produce the sequence
0111010000101000110101100111110101100000

is: print "0111010000101000110101100111110101100000"

If you think about it, you will see that as the random
number gets larger and larger, the size of the algorithm
which produces it must grow proportionately larger and
larger, whereas the algorithm to produce the non-random
number 1010101010... will stay roughly about the same
size, regardless of the size number it produces.

I first learned of this definition of randomness off of IBM
researcher Gregory Chaitin's website. He is the co-

discoverer of this concept

Another way of stating the above is that "algorithms CREATE ORDER or PATTERNS". Whenever you see a pattern, there is an underlying algorithm that has caused it. In mathematics, algorithms are called functions and in physics, algorithms are called natural laws. A function is an algorithm that relates one or more input values with a single output value.

For example:
$C = A * B$ is a function.
This function relates any two numbers A and B with a single product C.
It is also an algorithm. There is a step by step list of instructions you must follow in order to arrive at the answer C.

If you were given: $C = 5 * 6$, and did not know your multiplication tables, you would have to follow the instruction to "ADD 5 to itself 6 times".
Furthermore, addition itself can be broken down into an algorithm.

In physics, natural laws are written down as mathematical functions, Such as Force = Mass * Acceleration, and therefore natural laws are also algorithms. The point is that algorithms, mathematical functions, and natural laws are one and the same thing.

Patterns and Natural Law philosophy
Reality has an underlying order to it. This order is created by natural laws (the "rules" of physics). These natural laws apply to everything and everyone in the universe, whether they are aware of these laws or not. Natural laws are discovered through the use of <u>REASON</u>. Reasoning can be used to discover natural laws precisely because reality has

an underlying order to it. Natural laws are discovered by using your intelligence (ability to reason) to perceive and recognize patterns in nature. New mathematics is discovered by using your ability to reason to perceive and recognize patterns in numbers.

Furthermore, algorithms (natural laws and functions) COMPRESS information.
The string
10101010101010101010101010101010101010
can be compressed down to:
print "10" 20 times

There is a saying that "Theories destroy facts". What this saying means is that before a law of nature is known, you simply have LOTS and LOTS of data or information. Before Isaac Newton discovered the laws of gravitation, people like Kepler had produced detailed charts of the motions of planets across the sky. You could use these charts to help determine where the position of a planet should be. When newton discovered his law, you no longer needed all of the facts from the chart, all of the information contained in the chart had been compressed(reduced in size) into a simpler formula(algorithm).

Understanding IS Data Compression
You understand something when you can create an algorithm which can produce the pattern, event, or phenomenon you are trying to explain.

However, an algorithm that is as large as the the phenomenon you are trying to explain does not explain anything at all.

print "10" 20 times
is an algorithm that is 19 characters long(including spaces).

This algorithm explains the sequence of supposed coin tosses: 101010101010101010101010101010101010101010

An algorithm to explain nature reduces the information you see down to a smaller amount of information..."Theories(algorithms) Destroy Facts".

Phenomenon that CANNOT be explained in terms of a simpler algorithm, truly are RANDOM, in the since that they cannot be explained in terms of something simpler.

The size of an algorithm, and the size of the output it generates, can be measured in terms of INFORMATION. This brings up the question, What is information? Chapter 5 goes into detail about the mathematics of information.

Information and the Explanatory Filter
What is the difference between
01110100001010001101

and

10101010101010101010

They BOTH contain the SAME amount of information. The difference is that one of them can be generated by a simpler algorithm, the other cannot.

What is the difference between
"Hey fiddle diddle, the cat in the middle, the cow jumped over the moon" and
"sukachfulsad noa uhnah asiofjoihasd paisfdh aosidhjoaish roihas dfoiha"

They BOTH contain the SAME amount of information. Also NEITHER of them can be generated by a simpler algorithm.

According to the definition of Randomness I gave at the start of this article, they are BOTH random!

Yet intuitively, it does not seem possible that the "Hey fiddle diddle" string could be random. There seems to be a major difference between these 2 strings. What is this difference?

Imagine you have a friend go into another room and toss 70 scrabble pieces into the air. When your friend returns from the room, they tell you that when the scrabble pieces landed, they landed in this order: "Hey fiddle diddle, the cat in the middle, the cow jumped over the moon" Would you believe it? Probably not. Anyone would believe that your friend had intentionally laid the scrabble pieces onto the floor. The "Hey fiddle diddle" string fits a specified pattern of information that most people are familiar with, the old Mother Goose nursery rhyme. On the other hand, it is possible that when your friend tosses the scrabble pieces into the air, the letters "dog" may appear. Why, because the word "DOG" only contains 3 letters. It does not contain enough information in it to rule out the possibility of chance turning up the letters "DOG" in succession.

Whenever an event or phenomenon is sufficiently large in size (contains enough information), Cannot be generated by a simpler algorithm, and contains fits a SPECIFIED pattern, it can be concluded that the phenomenon or event can be attributed to an intelligent agent. Information that meets all 3 of these criteria is known as Complex Specified Information.

Chapter 3

Logarithms, The first calculators

The hand held calculator that everyone uses today first became available in the early 1970's. Prior to this time, if you wanted to do an arithmetic calculation, you had to do it by hand. Obviously, this takes A LOT of time, and so early on people began looking for ways to reduce the amount of work involved in multiplication and division.

In the late 16th century, a man named John Napier was one of those people searching for a way to make arithmetic calculations easier. At this time, the properties of powers (exponents) were known. Napier saw a PATTERN in the properties of the powers of numbers that led him to his discovery

What is a power?
Raising a number to a power simply means multiplying a number by itself.

For example,
2^2 reads "2 raised to the second power".
2^2 means 2 X 2 = 4

2^3 reads "2 raised to the third power".
2^3 means 2 X 2 X 2 = 8

2 is called the "base" and 3 is called the "exponent" or "power".

Here is a table of 2 raised to different powers.

	reads	means
2^1	"2 raised to the first power"	2
2^2	"2 raised to the second power"	2 X 2 = 4
2^3	"2 raised to the third power"	2 X 2 X 2 = 8
2^4	"2 raised to the fourth power"	2 X 2 X 2 X 2 = 16
2^5	"2 raised to the fifth power"	2 X 2 X 2 X 2 X 2 = 32
2^6	"2 raised to the sixth power"	2 X 2 X 2 X 2 X 2 X 2 = 64
2^7	"2 raised to the seventh power"	2 X 2 X 2 X 2 X 2 X 2 X 2 = 128
2^8	"2 raised to the eighth power"	2 X 2 X 2 X 2 X 2 X 2 X 2 X 2 = 256

If you think about it, you will realize that
if 2^2 = 2 X 2
and 2^3 = 2 X 2 X 2
then 2^2 X 2^3 = (2 X 2) X (2 X 2 X 2).

In other words, 2^2 X 2^3 = $2^{(2+3)}$ = 2^5

if you replace 2 with the letter A, then you will realize that
if A^2 = A X A
and A^3 = A X A X A
then A^2 X A^3 = (A X A) X (A X A X A).

In other words, A^2 X A^3 = $A^{(2+3)}$ = A^5

From looking for patterns in exponents, Napier realized that if you could rewrite all of the numbers in terms of a common base (base A, for instance), then you could reduce multiplication to addition, division to subtraction, powers to multiplication, and roots to division....

Napier's "invention" was literally to create a table of exponents as above. Only in this table, the exponent itself would be a separate column. Napier first wanted to name his table, the table of "artificial numbers", but later decided to call the table a table of "logarithms".

Here is a base 2 table of logarithms:

	Exponent	
2^1	1	2
2^2	2	4
2^3	3	8
2^4	4	16
2^5	5	32
2^6	6	64
2^7	7	128
2^8	8	256
2^9	9	512
2^{10}	10	1024

So how does this table make multiplication easier?

Say you wanted to multiply the numbers 64 and 16 together.
Looking at the entry for 64, you see that the exponent is 6.

Looking at the entry for 16, you see that the exponent is 4.

$6 + 4 = 10$
Looking down the exponent column for 10, you find the number 1024, which is the <u>ANSWER</u>!

By using this table, multiplication is reduced to addition. Also, division is reduced to subtraction, powers are reduced to multiplication, and roots to division.

In this table, the exponent column is also called the "logarithm" of the number in the right most column.

In other words,
$\log_2(64) = 6$
$\log_2(16) = 4$
$\log_2(1024) = 10$

Chapter 4

The relationship between the binomial expansion series, trees, probability, counting, and information

If you toss a coin into the air, it can land either heads or tails. If you write H for Heads and T for tails, you can draw the number of possible outcomes like this:

This diagram is called a "TREE". Each branch of the tree represents a different possible outcome from tossing a coin twice into the air. There are 2 possible outcomes from tossing a single coin into the air, and so there are 2 branches on this "Tree".

Below is a table summarizing the relationship between 1 coin toss, a tree, and the binomial expansion series from algebra:

1 coin toss	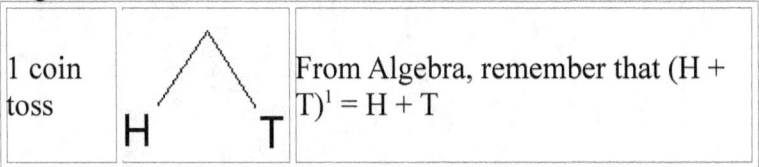	From Algebra, remember that $(H + T)^1 = H + T$

If you toss a coin into the air twice in a row, how many different possible ways can the coin land?

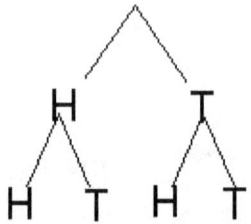

From Algebra, remember that
$(H + T)^2 = (H + T) * (H + T) = (H * H) + (H * T) + (T * H) + (T * T)$
and
$(H * H) + (H * T) + (T * H) + (T * T) = H^2 + 2HT + T^2$

Also notice that in algebra, HT "MEANS" H <u>times</u> T...So if you remove the * symbol you get:
$(H * H) + (H * T) + (T * H) + (T * T) = HH + HT + TH + TT$

A Pattern: Notice in the diagram below that
HH corresponds to the right most branch of the tree,
HT to the 2nd from the right branch of the tree,
TH to the 2nd from the left branch of the tree,
and TT to the left most branch of the tree

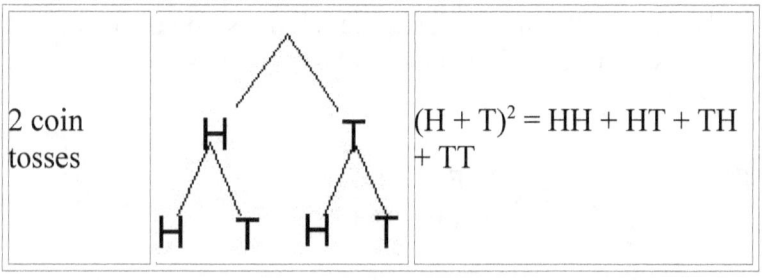

Each TERM in the binomial expansion represents a different branch on the tree.
Furthermore, you can see that $(H + T)^2$ physically MEANS

"tossing a coin into the air twice in a row, or tossing 2 coins into the air simultaneously".

After 3 coin tosses

$(H + T)^3$ = HHH + HHT + HTH + HTT + THH + THT + TTH TTT

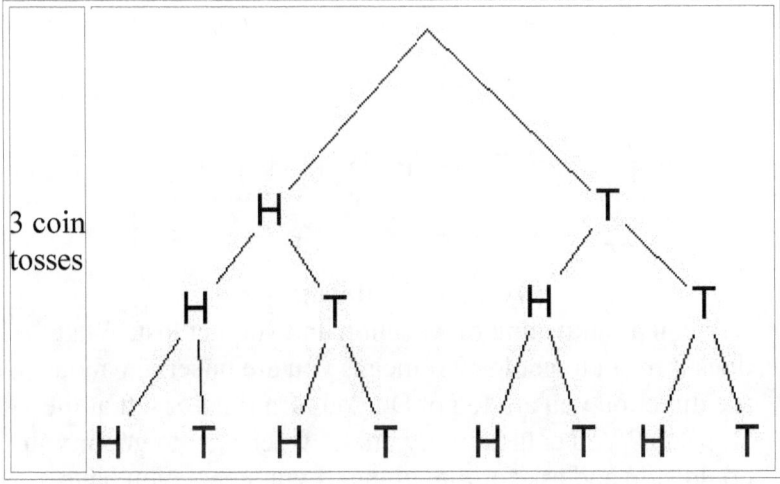

3 coin tosses

After 4 coin tosses

$(H + T)^4$ = HHHH + HHHT + HHTH + HHTT + HTHH + HTHT + HTTH + HTTT + THHH + THTH + THTT + TTHH + TTHT + TTTH + TTTT

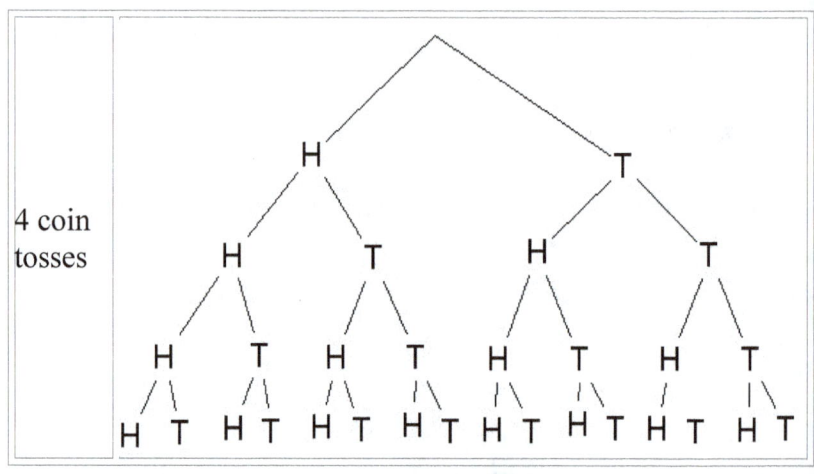

4 coin tosses

Trees and Information

Say you are traveling on vacation and you get lost. What does it mean to get lost? It means you are uncertain about the direction you are to go. Do you turn right or left at the stop sign? You could simply guess which way to go, or you could stop and ask for directions. If you guess, you have a 50% chance of being correct, or a probability of 1/2.

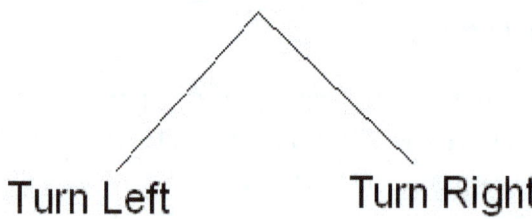

Turn Left   Turn Right

If you stop and ask someone if you should turn right or left, you will have acquired information. Similarly, if you guess that you should turn left and you discover that you have guessed incorrectly, you will have also acquired information (You now know that the correct direction to take at the traffic light is a right hand turn).

So, in some way, information is related to uncertainty (or probability), and acquiring information is to eliminate the uncertainty that you had (or ruling out possibilities).

This relationship was discovered by a man named Claude Shannon, and it is called INFORMATION THEORY. Chapter 5, What is Information, goes into detail about what probability and uncertainty have to do with the bits and bytes that are found inside of a computer!

The relationship between the binomial series, area, and volume
If you have a square with sides of length A,
The area of this square is $A * A = A^2$.

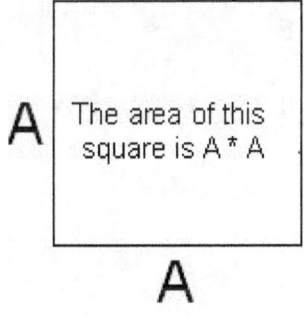

If you increase the length of each side of the square by a distance of B, so that each side of the square is now $A + B$, its new area is

$(A + B) * (A + B) = (A + B)^2$

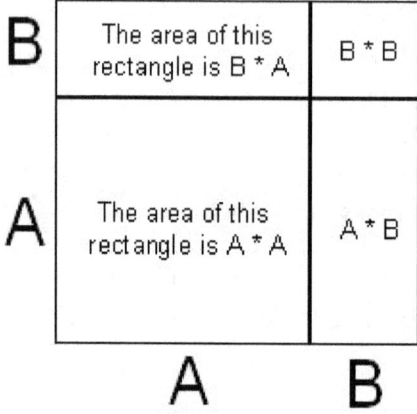

Adding the area of each of the above squares gives you the total area of the square whose sides are length A + B. This area is AA + AB + BA + BB

$$AA + AB + BA + BB = A^2 + 2AB + B^2$$

If you have a cube whose sides are all A, its volume is A * A * A = A^3

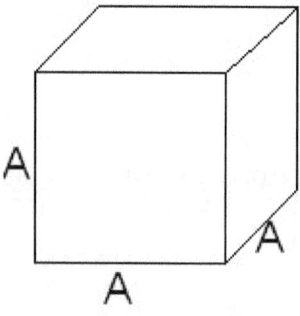

If you increase the length of each side of the cube by a distance of B, so that each side of the cube is now A + B, its new volume is (A + B) * (A + B) * (A + B) = $(A + B)^3$

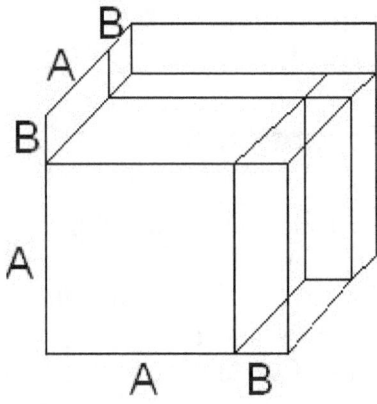

Can you see that the cube above is made up of 8 smaller cubes, and that the sum of the volume of these 8 smaller cubes EQUALS $(A + B) * (A + B) * (A + B) = (A + B)^3$

The binomial series can be thought of as both the tossing of a coin AND the surface area of a square or volume of a cube!!!

4 dimensional objects and hypercubes
Questions for further study:
If $(A + B)$ is the length of a line
and $(A + B)^2$ is the area of a square
and $(A + B)^3$ is the formula for the volume of a cube,
then does $(A + B)^4$ have any physical, geometric interpretation? What about $(A + B)^5$, $(A + B)^6$ or higher? Are these the volumes of hypercubes?

Chapter 5

What is information?

Say you are traveling on vacation and you get lost. What does it mean to get lost? It means you are uncertain about the direction you are to go. Do you turn right or left at the stop sign? This decision can be represented by a graph called a TREE. Below is a picture of a tree representing the decision to turn right or left.

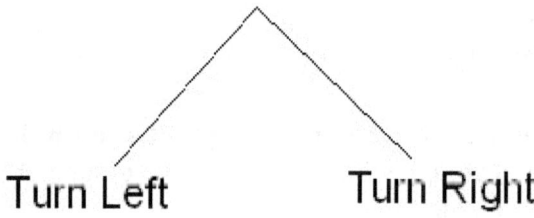

You could simply guess which way to go or you could stop and ask for directions. If you guess, you have a 50% chance of being correct, or a probability of 1/2.

You can see that you have exactly 2 choices at the intersection... Turn Left or Turn Right. Only 1 of these 2 choices is the correct one, so you have a probability of 1/2 of making the correct choice. (To learn more about trees and probability, see chapter 4.)

If you stop and ask someone if you should turn right or left, you will have acquired information. Similarly, if you guess that you should turn left and you discover that you have guessed incorrectly, you will have also acquired information (You now know that the correct direction to take at the traffic light is a right hand turn).

So, in some way, information is related to uncertainty (or probability), and acquiring information is to eliminate the uncertainty that you had (or ruling out possibilities).

If you stop and ask someone for instructions, and they tell you more than just turn right or left at an intersection, but give you a complete, turn by turn set of instructions through a series of 5 intersections, then you will have acquired even more information still.

If at each of these intersections, you can only turn right or left, then there is a probability of 1/2 of guessing the right direction to go at each intersection. This corresponds to $(1/2)*(1/2)*(1/2)*(1/2)*(1/2) = 1/32$. There is a probability of only 1/32 of guessing your way correctly to the destination.

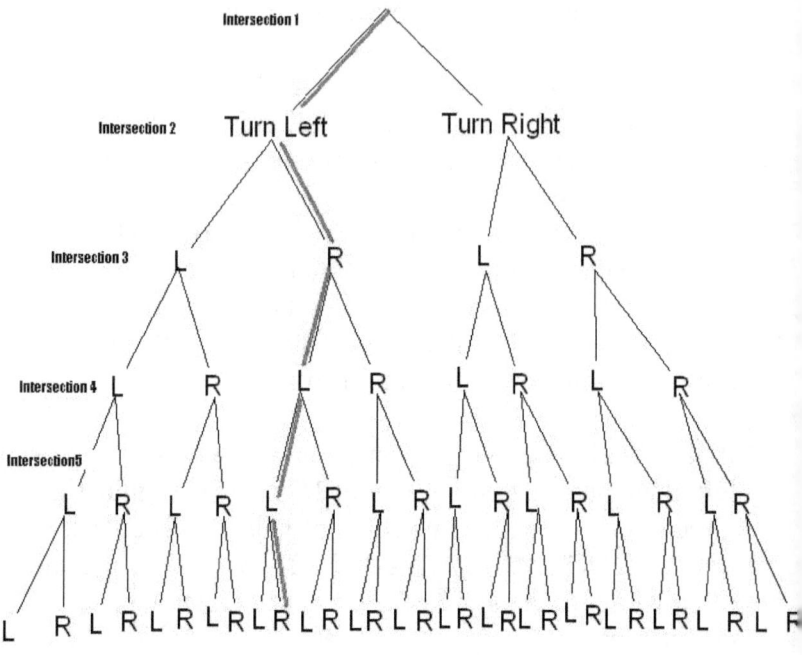

(Highlighted in red is one of the 32 possible choices you can make if you have to guess your way through 5 intersections. The highlighted branch corresponds to Turning Left at the first intersection, Right at the second intersection, Left at the third, Left at the fourth, and Right at the fifth.)

This means that if you HAD to guess your way through 5 intersections, you would be only 3.1% certain that you would arrive at your destination. On the other hand, you had a certainty of 50% that you would guess your way to your destination correctly if you only had to guess at a single intersection, as in the first example.

So, the less certainty you have about an event or situation (such as how to get to where you need to go), the more information you receive when you get an answer. You acquired MORE information when you were given turn by turn instructions through 5 intersections (corresponding to only 3.1% certainty) as opposed to when you only needed to know which way to turn at ONE intersection (corresponding to 50% certainty).

A question you may have at this point is, what does probability and uncertainty have to do with the bits and bytes that is used to measure information content and storage inside of a computer? The answer was discovered in the early part of the 20th century buy a man named Claude Shannon.

How to measure (quantify) information
We live in the information age, and it is pretty common knowledge these days that computers are made up of patterns of nothing but 1's and 0's. The smallest unit of information is called a bit. People buy hard drives measure in terms of Giga Bytes, or billions of bits. Conceptually, a single bit can be thought of as one light bulb.

When the light bulb is on, it represents the number 1. When the light is off, it represents the number 0.

So a single bit can be a 1 or a 0.

Likewise, 1 quarter can also represent 1 bit of information...either a 0 or a 1.

For example, heads can represent 1
and tails can represent 0

How does a quarter tie in with the example I gave at the beginning of this article about information relating to uncertainty?? Here it is...

The mathematical relationship between the uncertainty or probability of an event occurring (which I will label, P) and the information specified by the event (which I will call I) is : $I = -\log_2(P)$

(Note: log stands for "logarithm")

If you do not know what a logarithm is, see chapter 3, Logarithms, the first calculators. (Historically, logarithms were actually the FIRST calculators!)

What does this mean?

I just said that a quarter contains 1 bit of information. Likewise, if you toss a quarter in the air, you have a 50% chance of it landing heads or tails...or probability $P=1/2$

Plugging 1/2 or .50(50%) into the above formula yields

$I = -\log_2(1/2)$ $I = -(-1) = 1$ bit!

Mathematically, an event of uncertainty of 50% corresponds to 1 bit! This is why the bit is the smallest unit of information. It literally represents the simplest possible choice that you can make. You can either choose option A or option B...you have no other choices, and if you were to remove either of the options then you would have no choice or uncertainty at all, and hence <u>no</u> information!

What is the probability that tossing 3 quarters in the air will turn up heads each time, and how much information is specified by this event?

There is a probability of 1/2 of a coin landing heads once, and a probability of $(1/2)*(1/2)*(1/2) = 1/8$ of it landing heads 3 times in a row.

So P=1/8
and
$I = \log_2(1/8)$
$I = - (-3) = 3$ bits.

So 3 bits corresponds to 3 quarters!

This mathematical relationship between uncertainty and information was discovered by one of the pioneers of computer science, Claude Shannon, and is called Shannon Information in his honor.

Chapter 6

Information that is the product of intelligence

This book has introduced the idea that wherever a pattern is found, there is an underlying algorithm or law of some sort that has created it.

The information found in the universe can be the result of 3 things:
1. It can be the product of an underlying algorithm or law, which has been the theme of this book thus far.
2. It can be the product of pure chance…a purely random collection of information.
3. It can be the product of intelligence …the result of a mind.

The mathematician William Dembski has created a statistical filter used to determine which of the 3 possible causes any information found in the universe is a product of. He calls this filter, the explanatory filter. It asks 3 questions:

1. Does the information fit a pattern that can be described in terms of a simpler underlying law or algorithm? If it does, then the information is a **necessary** consequence of the law or algorithm that created it. If it does not, then the information is **contingent**. The word contingent means something that is not necessary, something that is subject to chance or unforeseen causes.

2. If the information is not the product of an underlying law or algorithm, how **complex** is the information? The

34

sequence Heads, Tails, Tails, Tails, Heads, Tails can easily be explained as the product of six random coin tosses. It is the product of **chance**. In Dembski's explanatory filter, if an event has a likelihood of occurring of at least 1 in 10^{150}, it can safely be attributed to chance. An event of probability 1 in 10^{150} corresponds to 500 bits of information…500 coin tosses. Information that contains less than 500 bits of information and cannot be described in terms of a simpler underlying algorithm or law can be attributed to **chance**. It is a random event. (From Chapter 5, we know that the information contained in an event of with a probability of occurring of 1 in 10^{150} is: $-\log_2(10^{-150}) = 500$ bits)

3. Does the information match a **specific** pattern? Information that matches a specified pattern, and is sufficiently complex enough, containing at least 500 bits of information, can be safely attributed as the product of an intelligent designer. This type of information is called **complex specified information**. If the information does not match a specified pattern, and cannot be described in terms of a simpler algorithm or laws, but contains 500 bits of information or more, it is a product of chance.

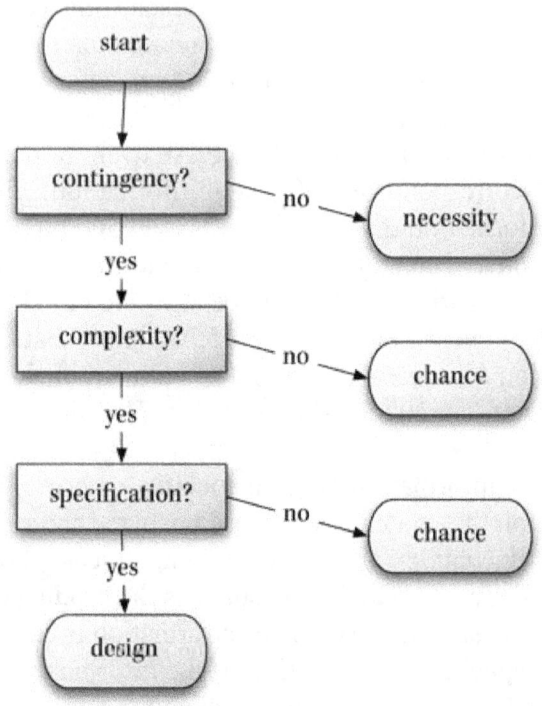

Above, William Dembski's Explanatory Filter

William Dembski often uses SETI, the search for extraterrestrial intelligence, as an example of using the explanatory filter to determine the origin or cause of information. SETI uses radio telescopes to monitor radio waves from outer space as part of its search for extraterrestrial life. The information contained in these radio waves is sent scanned by computers programmed to look for patterns in this information. How will SETI know if they have received a radio signal coming from an alien intelligence? The short answer is they must use some variation of the explanatory filter.

In the movie Contact, based upon a book by Carl Sagan about a group of SETI researchers, a radio signal coming from extraterrestrial intelligence was found. The researchers in this movie received the radio signal as a sequence of 1126 beats and pauses, where 1s correspond to beats and 0s to pauses. This formed a string of 1s and 0s similar to the string of 1s and zeros recorded for coin tosses in chapter 2. Another name for a string of 1s and 0s is a bit string. The bit string in the radio signal from the movie contact was the binary representation of the prime numbers from 2 to 101. Each beat in the sequence corresponded to a bit...for a total of 1126 bits of information.

Such a signal would meet all of the criteria for inferring information caused by an intelligent designer, a mind. The information could not be described in terms of the simpler laws of physics responsible for the creation of the radio waves. The sequence of beats corresponded to 1126 bits of information, far more than the 500 bits needed to ensure the information was complex enough to rule out chance. Finally, the information in the radio signal matched a specified pattern of information...the prime numbers from 2 to 101.

Other examples of agencies that employ a variation of the explanatory filter are crime scene investigators. Was the scene being investigated the result of chance, or is there sufficient evidence to indicate that the scene was master minded with ill intent?

Complex specified information is found throughout the world we live in. Art and engineering...all of this of course contains complex specified information.

Mount Rushmore, below, is an example of Complex Specified Information.

Chapter 7

Simple Decision Making Machines

What Is A Computer?

This book has introduced the idea that the underlying order of the world around us, the patterns that we see, are the products of laws or rules of behavior, also called algorithms.

A computer is a machine that follows a step by step list of instructions. Another name for a step by step list of instructions is an algorithm. Examples of algorithms people encounter on a day to day basis are a list of items to buy at a grocery store, the instructions to assemble a bicycle, and the rules that everyone is supposed to follow when they come to a traffic light. The purpose of all computers, the reason for their existence, is to follow algorithms. Another name for the algorithm a computer follows is a computer program.

In a sense, a computer follows the instructions in an algorithm in exactly the same way that a person follows the rules of a board game such as Monopoly. The rules of a board game can be thought of as the computer itself. A board game has a series of squares on it. These squares are just like a computer's memory. Each square in a board game typically holds an instruction, such as "Go to jail", "Collect $200.00", etc.
These instructions are analogous to the individual instructions making up a computer program / algorithm.

(A computer follows the instructions in an algorithm in exactly the same way that a person follows the rules of a board game such as Monopoly.)

In a board game, players move from one square to the next. When a player moves to a new square, they perform the instruction found on this square. Likewise, a computer moves from one of its "memory squares" to the next. When a computer moves to a new memory square, it performs the instruction on the square it moved to.

In a board game, when a player moves to a new square, they typically have to make some sort of decision based upon the condition of the square they landed on. For example, in Monopoly, IF you land on a square containing a property, and IF the property is not yet owned, THEN you may buy the property. This changes the state of the square from representing an un-owned property to representing an owned property. IF the square you move to contains a property that is owned, THEN you must pay a fee. Other squares may instruct you to JUMP to a new square, skipping over all the squares in between, such as the "Go to jail" square.

This is not just a good analogy of how a computer works. Historically, this is how the computer was first conceived. In the 1936 paper titled "ON COMPUTABLE NUMBERS, WITH AN APPLICATION TO THE ENTSCHEIDUNGSPROBLEM" Alan Turing, the inventor of the computer, described the computer in terms of a step by step list of rules to follow, similar to the rules of a board game. His theoretical computer would later be named a "Turing Machine" in his honor. The Turing machine could move along a series of squares, just like the squares described above. Based upon the condition of the square the computer moved to, it could do one of 3 possible things. It could do nothing, change the condition of the square, or move to a new square.

These features embody the essence of all computers. In a nutshell, the simplest possible computer is one that can

1. move from one square to the next
2. read the state of the square it moves to
3. make a decision based upon the state of the square.

In step 3, the computer can decide to take one of three possible courses of action. It can decide to do nothing, change the state of the square it is on or any other square on the board, or jump to any other square.

By the end of this chapter, you will understand in detail how switches are cleverly combined together to implement step 3…automating the task of making a decision.

Searching for Machines that "Make Decisions"

What is a decision? A decision is a choice between 2 or more possibilities. For example, if you are lost while driving down a one way road and come to a fork in the road, you must make a decision. You must decide whether to turn right or left. The simplest decision you can make is one where a MINIMUM of 2 possible choices are involved (Simple in terms of number of possibilities to choose from, not in terms of the moral or ethical dimensions of a decision).

Consider the decision, "Do I go fishing?" To make the decision, you must do several things. You must:

1) Decide what possible courses of action to be taken. In this example there are two, either
      1. I will go fishing
   Or 2. I will NOT go fishing.

2) Decide the event or thing that will determine which course of action will be selected. In this example, the event I have selected is "I feel like it". "I feel like it" is a proposition. **A proposition a statement that is either true or false.**

A feature of ALL decisions is that the outcome of the event or thing that determines which course of action will be taken can ALWAYS be written in the form of a proposition.

3) Map each possible action listed in step 1 to one of the event's or proposition's outcomes listed in step 2. These mappings can take the form of an "IF – THEN" statement. In this example, there are two mappings:

     1. IF (I feel like it) THEN (I will go gishing)

And 2. IF (I do not feel like it) THEN (I will not go fishing)

All IF – THEN statements take the form of: IF (This event / proposition is true) THEN (perform this action)

More than one action can be mapped to a single outcome of an event / proposition. For example, IF (I feel like it) THEN (I will go fishing) AND (I will go swimming).

A switch, the simplest decision automation machine
A switch, like the switch you use to turn a lamp on or off, is a natural candidate for a machine that can automate making a decision. A switch, like a proposition, has two states. A proposition can be either true or false, and a switch can be either open or closed, on or off. You could just as easily label the on and off states of a switch as "True" and "False", respectively.

Below is an example of what must be done to automate the decision, "Do I sound the alarm?" with a switch. To automate this decision, you must do several things:

1) You must decide what possible courses of action can be taken. In this example there are two, either
    1. Sound the alarm
  Or 2. Do not sound the alarm

2) You must decide the event or thing that will determine which course of action will be selected. This can always be written as a proposition. In this example, the proposition I have selected is "The window is open". All propositions can be either true or false, and so this proposition has two possible outcomes:
    1. The window is open
    2. The window is not open

3) You must map each possible action listed in step 1 to one of the event's or proposition's outcomes listed in step 2. These mappings can take the form of an "IF – THEN" statement. In this example, there are two mappings:

        1. IF (the window is open) THEN (sound the alarm)

    And 2. IF (the window is not open) THEN (do not sound the alarm)

Below is a diagram of a switch that automates this decision. The switch is permanently kept open as long as the window is closed. When the window is opened, a spring beneath the switch pushes the switch closed, closing the electrical circuit and allowing electrical energy to travel through the bell, sounding the alarm.

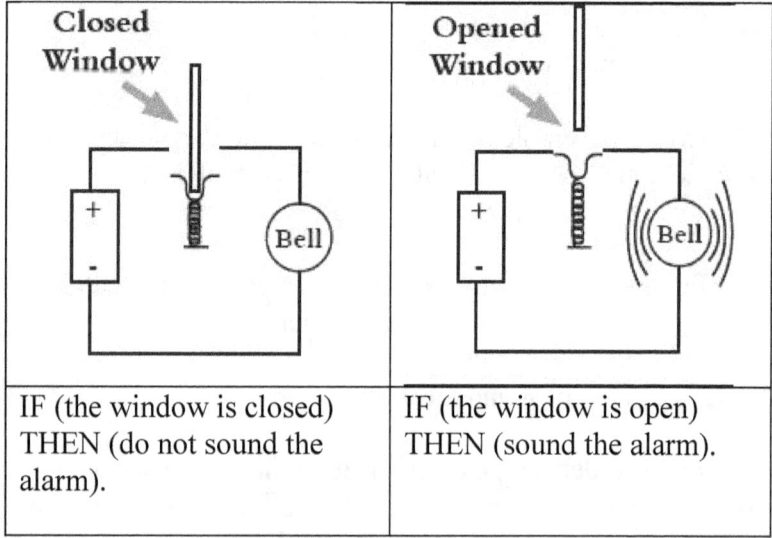

Closed Window	Opened Window
IF (the window is closed) THEN (do not sound the alarm).	IF (the window is open) THEN (sound the alarm).

The window alarm switch in this example does not make any decisions (nor does any other switch). All decisions concerning when the window is considered open and what action should be taken if the window is open were decided in advance by a pre-existing mind, the mind that designed the switch.

While this is intuitively obvious in the case of a single electric switch, it is far less obvious in the case of a sophisticated computer or a brain, both of which contain millions or billions of switches (millions in the case of computers and billions in the case of human brains).

The number of switches involved, or the way they are connected together, does not change the fact that minds, and not switches, make decisions.

The term "decision making machine" is misleading. The term "decision automation machine" more accurately describes the machine's function…that of automatically carrying out a decision that was made ahead of time by the mind that designed/built the decision automation machine/switch.

The scientist Claude Shannon was the first person in the world to recognize that propositions can be physically represented as electrical switches.

With this insight, in his 1938 paper, "A Symbolic Analysis of Relay Switching Circuits", Claude Shannon showed how the branch of mathematics called Propositional Logic could be completely represented by electrical switches. (This paper, and the paper by Alan Turing referenced above, are the two founding papers for ALL of modern computer engineering and computer science).

Propositional Logic is the study of the logical relationships between 2 or more propositions. In his 1858 book titled "The Laws of Thought", George Boole showed that there are only 3 possible relationships between any two propositions, from which propositional statements of greater and greater complexity can be derived. These are the AND relationship, the OR relationship, and the NOT relationship. On page 11 of his paper, Shannon shows how each of these relationships can be created by connecting switches(propositions) together in series or in parallel.

A circuit containing a single switch that is closed (also called a closed circuit) represents a TRUE proposition.

A circuit containing a single switch that is open (also called an open circuit) represents a false proposition.

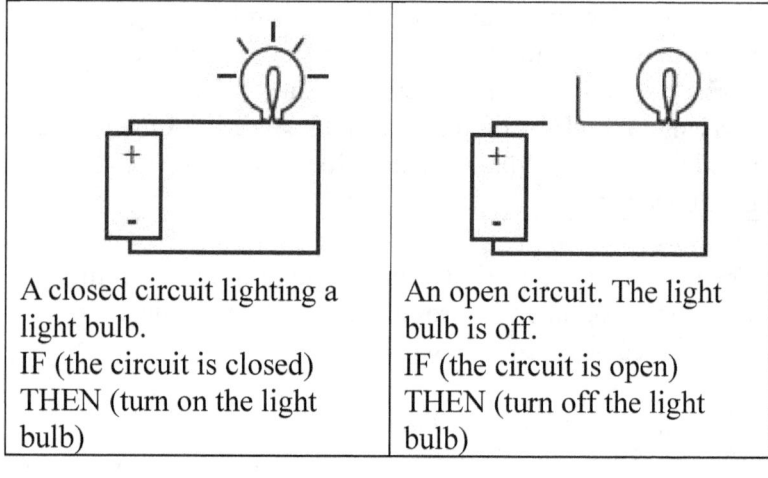

A closed circuit lighting a light bulb. IF (the circuit is closed) THEN (turn on the light bulb)	An open circuit. The light bulb is off. IF (the circuit is open) THEN (turn off the light bulb)

The AND relationship is represented by two or more switches connected together in series:
For example,
IF (the Garage Door is open)
AND IF (The Close button is pressed)
THEN (Turn on the electric motor that closes the garage door (the down motor))

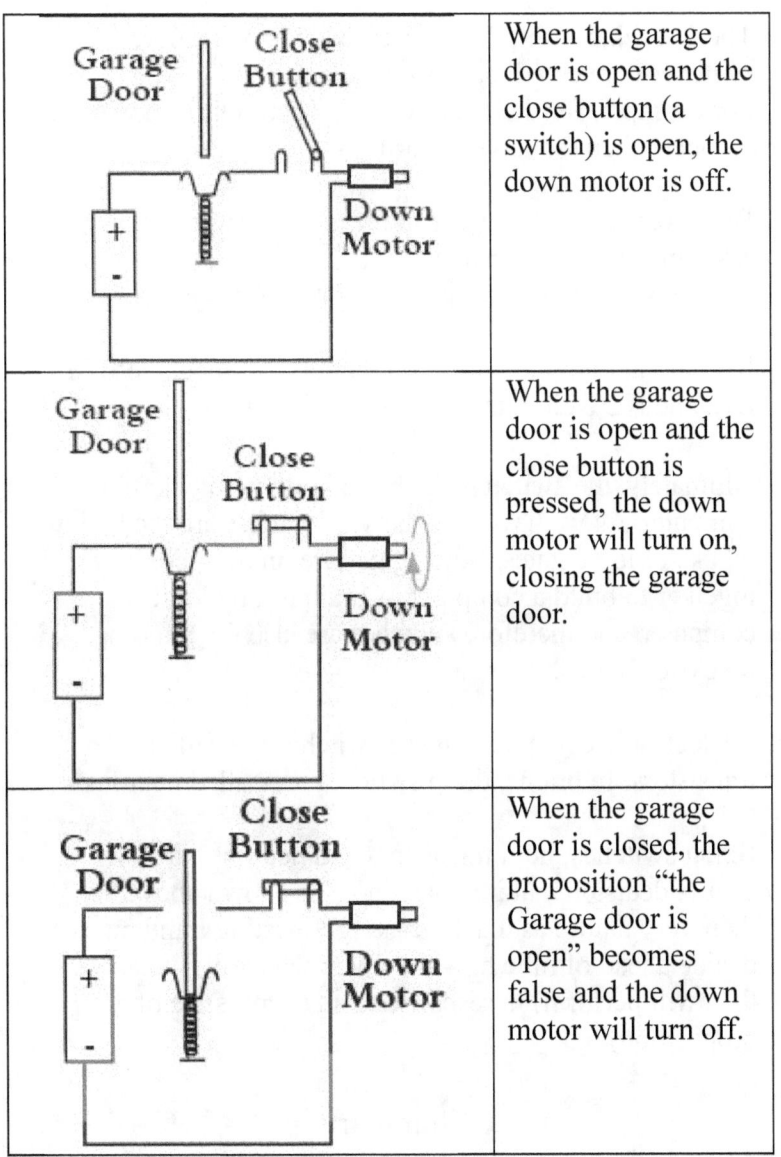

	When the garage door is open and the close button (a switch) is open, the down motor is off.
	When the garage door is open and the close button is pressed, the down motor will turn on, closing the garage door.
	When the garage door is closed, the proposition "the Garage door is open" becomes false and the down motor will turn off.

The OR relationship is represented by two or more switches connected together in parallel:
For example, IF (the window #1 is open) OR (the window #2 open) then (sound the alarm).

By repositioning the wires in the alarm switch, The NOT relationship can be created:
IF (the window is **not** open) THEN (sound the alarm)

In computer engineering and computer science, these AND, OR, and NOT circuits are called logic gates.

Ultimately, the switch is the basic building block of a computer. As we have just shown, switches are the building blocks of logic gates. Logic gates are, in turn, connected together to build a computer. A brain is a biological computer, and therefore switches are also the building blocks of brains.

In electronic computers, these switches are called transistors. In brains, these switches are called neurons.

Being switches, both transistors and neurons act as IF – THEN decision making machines. As shown above, by their very nature as machines, a pre-existing mind must decide ahead of time how they will behave(what action they will perform) in response to an event (state of a proposition).

Summary

At each and every step of the way, a pre-existing mind is required to decide ahead of time how a decision automation machine should behave…the course of action that should be taken in response to a certain event. Minds, not machines, make decisions.

This rule continues to apply when individual decision automation machines (switches) are connected together to form logic gates, and these logic gates are connected together to form logic circuits. This rule applies no matter how many switches are connected together, and no matter how complicated the arrangements of these connections become.

From switch, to logic gate, to computer…a pre-existing mind is the cause of all apparent decision making and behavior in these machines…in ANY machine, including your brain (which is a machine).

This rule applies to the individual neurons (biological switches) comprising your brain, as well as to neurons connected together to form a network. The rule applies no matter how many neurons there are or how complicated the arrangement of the connections in the neural network become.

Mind is not the product of a computer or brain, the opposite is true. Computers and brains are the product of pre-existing minds.

www.ingramcontent.com/pod-product-compliance
Lightning Source LLC
Chambersburg PA
CBHW051254170526
45165CB00004B/1705